BIOTECHNOLOGY AND AGRICULTURE IN 2020

T0130700

A Report of the Project on Technology Futures and Global Power, Wealth, and Conflict

Project Director
Anne G.K. Solomon

Author
Anthony J. Cavalieri

April 2005

About CSIS

The Center for Strategic and International Studies (CSIS) is a nonprofit, bipartisan public policy organization established in 1962 to provide strategic insights and practical policy solutions to decisionmakers concerned with global security. Over the years, it has grown to be one of the largest organizations of its kind, with a staff of some 200 employees, including more than 120 analysts working to address the changing dynamics of international security across the globe.

CSIS is organized around three broad program areas, which together enable it to offer truly integrated insights and solutions to the challenges of global security. First, CSIS addresses the new drivers of global security, with programs on the international financial and economic system, foreign assistance, energy security, technology, biotechnology, demographic change, the HIV/AIDS pandemic, and governance. Second, CSIS also possesses one of America's most comprehensive programs on U.S. and international security, proposing reforms to U.S. defense organization, policy, force structure, and its industrial and technology base and offering solutions to the challenges of proliferation, transnational terrorism, homeland security, and post-conflict reconstruction. Third, CSIS is the only institution of its kind with resident experts on all the world's major populated geographic regions.

CSIS was founded four decades ago by David M. Abshire and Admiral Arleigh Burke. Former U.S. senator Sam Nunn became chairman of the CSIS Board of Trustees in 1999, and since April 2000, John J. Hamre has led CSIS as president and chief executive officer.

Headquartered in downtown Washington, D.C., CSIS is a private, tax-exempt, 501(c) 3 institution. CSIS does not take specific policy positions; accordingly, all views expressed herein should be understood to be solely those of the author(s).

Library of Congress Cataloging-in-Publication Data
Cavalieri, Anthony J.
Biotechnology and agriculture in 2020 : a report of the project on technology futures and global power, wealth, and conflict / author, Anthony J. Cavalieri.
p. cm.
ISBN 0-89206-461-7 (alk. paper)
1. Plant biotechnology. I. Project on Technology Futures and Global Power, Wealth, and Conflict. II. Center for Strategic and International Studies (Washington, D.C.) III. Title.
SB106.B56C38 2005
631.5'233—dc22

2004031098

© 2005 by the Center for Strategic and International Studies. All rights reserved.

The CSIS Press
Center for Strategic and International Studies
1800 K Street, N.W., Washington, D.C. 20006
Tel: (202) 887-0200
Fax: (202) 775-3199
E-mail: books@csis.org
Web: http://www.csis.org/

Contents

Preface

Throughout 2004, at the request of the National Intelligence Council (NIC),[1] the Center for Strategic and International Studies (CSIS) brought together leaders from advanced technology firms, venture capital enterprises, research universities, and government to consider the geopolitical, economic, and social implications of technological advance out to the year 2020. This publication is one of a series that reflect their deliberations.[2]

The purpose of the exercise was to provide the NIC with ideas and insights relevant to NIC 2020, a project designed to provide U.S. policymakers with a view of how world developments could evolve and to identify opportunities and potentially negative developments that might warrant policy action. The challenge for CSIS was to produce for the NIC analyses that would be useful over time in the face of an explosion of scientific and technological knowledge and with great uncertainties in their paths of growth and patterns of interaction with economic, social, and other forces.

CSIS work included a series of workshops, conferences, and commissioned papers that considered frontier research and innovation and related policy challenges: the global ubiquity of powerful, dual-use technologies; the increasing diversity of players in the global advanced technology enterprise that operate beyond the influence of governments or international governing bodies; and the policy dilemma of balancing security-driven imperatives to curtail the availability of dual-use assets with the need of advanced technology enterprise for openness.

Numerous individuals provided advice and assistance throughout the project. Adrianne George, CSIS program coordinator, provided extensive intellectual and administrative assistance. We owe special debt of gratitude to Dr. Charles A. Sanders, CSIS trustee and project chair, and to Frank C, Carlucci, honorary cochair. CSIS would also like to thank Ambassador Robert L. Hutchings, former chairman of the National Intelligence Council, for his encouragement and support.

Anne G.K. Solomon
Project Director

---

[1] The NIC is a center of mid-term and long-term strategic thinking within the U.S. government, reporting to the director of central intelligence and providing the president and senior policymakers with analyses of foreign policy issues.

[2] Sandra Braman, *Information Technology, National Identity, and Social Cohesion* (CSIS, 2005); Anthony J. Cavalieri, *Biotechnology and Agriculture in 2020* (CSIS, 2005); Gerald L. Epstein, *Global Evolution of Dual-Use Biotechnology* (CSIS, 2005); Julie E. Fischer, *Dual-Use Technologies: Inexorable Progress, Inseparable Peril* (CSIS, 2005); David Nagel, *Wireless Sensor Systems and Networks: Technologies, Applications, Implications, and Impacts* (CSIS, 2005); and Anne G.K. Solomon, ed., *Technology Futures and Global Power, Wealth, and Conflict* (CSIS, 2005).

Biotechnology and Agriculture in 2020

Anthony J. Cavalieri
Senior Associate
Center for Strategic and International Studies

This report was commissioned by the Center for Strategic and International Studies as part of the "Technology Futures and Global Power, Wealth, and Conflict" project supported by the National Intelligence Council.

Executive Summary

The ability of countries to feed their people depends on natural resources, population, and technology for producing food. The earth's human population is expected to level off at about 9 billion people in 2050. Most good agricultural land is already in production, so additional food will have to come from higher productivity on existing farmland. Many existing technologies—plant breeding, hybridization, pest control, fertilizer, cropping systems, irrigation, and post-harvest storage technologies—can contribute to increased food production if applied more broadly. Agricultural biotechnology also offers the possibility of higher productivity and new agricultural products.

More than 67 million hectares of transgenic crops were grown in 2003. Traits of value to farmers will soon be joined by traits of value to consumers and industry. Basic scientific developments in genomics are facilitating the discovery of genetic traits that are valuable in agriculture. Research spending by companies participating in the seed business and public research organizations has moved these discoveries toward product introduction.

The United States leads in research spending and in the adoption of transgenic crops. China, India, and Brazil are developing meaningful research capacity in agricultural biotechnology and could dominate the field by 2020. Capacity to apply biotechnology in developing countries in Africa and elsewhere is largely missing despite the promise of reducing hunger. Many challenges stand in the way of applying agricultural biotechnology broadly. In order to ensure environmental and food safety, expensive regulatory systems have been developed. This cost hurdle prevents all but the most profitable commercial traits and crops from finding their way into farmers' hands. In Europe, commercialization of agricultural biotechnology has been blocked. Finding a balance between ensuring the safety of transgenic crops and the cost and complexity of the regulatory system will be essential for broader use of biotechnology. The ability to patent genes, crop varieties, and technologies further complicates the broad application of agricultural biotechnology, particularly in the developing world. However, a number of approaches will allow broad public access to important technology by 2020.

Large differences in the adoption of transgenic crops will continue as countries determine their regulatory requirements for introduction. Should the United States, European Union (EU), and Japan carry through with their promises to reduce agricultural subsides, competition among farmers throughout the world may become a reality. Use of new technologies may well provide a competitive advantage in such a world. While application of agricultural biotechnologies to terrorists' ends is possible, it seems to be a relatively low-level threat.

Production of crops on increasingly marginal land represents a significant threat to biodiversity and the environment. Technologies that increase productivity of existing farmland can prevent loss of natural habitat. Agricultural biotechnology may also decrease the environmental impact of traditional agriculture.

Advances in basic science will drive the development of agricultural biotechnology. Regulatory systems and public acceptance will determine the extent to which the technology is actually applied. Proper management could result in benefits to the environment and the ability to feed the world's population. Unless significant additional research funding becomes available, it is unlikely that agricultural biotechnology will help feed the hungry people of the world by 2050.

Introduction

The application of biotechnology to crop plants will have a major impact on global agriculture in the future. Advances in science will make the manipulation of plant genetics increasingly easy and more cost effective. The amount of time between now and 2020 is short when compared to the time required for the invention and implementation of new technologies. This paper will identify important agricultural technology trends in crop plants and speculate on their social, political, and economic implications between now and 2020, and beyond.

It should be recognized that the full implications of the adoption of many of these technologies could occur after 2020.

Global agriculture takes many forms, from large technologically sophisticated, industrial-scale farms to small subsistence plots. Efficiencies of scale allow mechanized farms in developed countries to produce large amounts of cheap food. The importance of farmers as a political constituency in the developed world has driven governments to provide large subsidies for smaller and smaller numbers of farmers who are able to grow adequate, and in some cases surplus, supplies of food. In the developing world, urbanization, HIV/AIDS, lack of production and distribution infrastructure, poor-quality farmland, inadequate water, and rapidly increasing populations make the production of adequate food more challenging. These disparate types of agriculture make generalizations difficult, thus it is important to draw different conclusions for different world regions.

First and foremost, the goal of global agriculture is to feed the world's population. The earth's current population is approximately 6.3 billion people; 842 million of these people do not have enough to eat. By 2020, the earth's population will have increased by at least 1.2 billion, to 7.5 billion people. Almost all of this population growth will occur in the developing world where agriculture is less productive. In addition to a more rapidly growing population, increased personal income, particularly in Asia, will result in additional meat consumption, increasing the need to produce more grain for animal feed. The earth's population is expected to level off at around 8.9 billion people in 2050 (UN Population Division 2003). Assumptions about population trends are critical in projecting the need for food production. The leveling off of human population growth is among the most important of these assumptions. When or if population stability is reached has a profound impact on the requirements for agricultural production. It is possible to make predictions of the need for food based on population growth by 2020, the focus of this paper, and to draw conclusions about the impact of the application of biotechnology to food production within this time period. It is more difficult to predict population growth farther into the future, a time when many of the biotechnologies discussed in this paper will be more mature and available for broader application.

Almost all of the worldwide increase in cereal production since 1960 has come from increased yields rather than increased land under production. These increases have come from a combination of improvements in crop genetics and management (Runge et al. 2003). Available farmland is not thought to be limiting to production of adequate food for the population through 2050. About 1500 m^2 per capita of land are required for production of an adequate diet. Farmland under cultivation is predicted to increase by about 18 percent by 2050 (Smil 2000). Even though the world has adequate land to produce food, productivity on that land will need to increase. Strong research performance will be required to raise productivity and minimize the effects on the environment (Runge et al. 2003). Incorporation of all appropriate technological approaches will increase the likelihood of achieving the goal of adequate food.

Agriculture is a significant factor in the global economy. Many developing-country farmers who produce marketable surpluses can earn income and the possibility for export. At the same time, agriculture no longer represents a large part of the economies of most developed countries. Technology capable of impacting both developing and developed countries raises an array of issues about the social and economic impact of the adoption of these technologies Agricultural issues are sometimes addressed in a way that reflects short-term political pressures, even though carefully thought out, long-term solutions are needed. Population increases and the vagaries of climate, water, and soil quality create an underlying fragility of food production. Food production is also susceptible to changes in climate in the most productive areas. Since crop pests such as insects, bacteria, and fungi are changing all the time, it is impossible to predict when a more virulent disease or more voracious insect might arise and spread. It is always possible that new forms of resistance will not exist or be readily available. The existence of these contingencies requires an ongoing, in-depth examination of the impact of changes in agricultural production on the food supply and world economy.

It is unlikely that many nations will be able to grow all of their own food in the future. Increased trade will be necessary to provide variety and quantity of food (Runge et al. 2003). In spite of the increase in trade, supplying the world's people with food is not merely a distribution problem. In recent years, total global food production has been adequate to feed the earth's population. Famine and hunger have resulted from locally reduced food production and poor distribution caused by wars, ineffective government policies, and natural disasters such as flood or drought. These types of food crises can be solved in the short term by food aid and redistribution. However, sustained adequate global food production cannot be achieved through improving distribution systems alone, since many countries are not rich enough to buy food on the world market.

Pinstrup-Andersen and Schioler (2001) have addressed the need for local food production. They state that agriculture is a major source of income for farmers and others living in surrounding communities. Self-sufficiency followed by excess agricultural production allows smallholder farmers to earn additional income. Redistribution of food from the world's rich countries weakens poor countries' local capability and initiative and eventually their opportunity for export. As Pinstrup-Andersen and Schioler (2001) point out, "in almost all developing countries...agriculture is the predominant industry and the most important source of income for the poor. A large slice of GNP is derived from agriculture, whether through production or processing, since agro industry is often one of the few reasonably well-established industries in poor countries."

Agriculture has the largest effect on land use and biodiversity of any of man's activities. Implementation of sustainable agricultural systems using an appropriate mix of existing and new technologies can reduce the impact of food production on the earth's resources and offer the promise of feeding the population on a smaller land area with fewer pesticides and other harmful inputs.

Conventional Agricultural Technologies

Although the focus of this analysis is on new agricultural technologies, conventional technologies will be briefly examined. Many proven technologies, which increase crop production and reduce the environmental consequences of agriculture, are currently in use to varying degrees around the world. Crop productivity has increased over the last 100 years through improvements in both plant breeding and management. Cereal-grain yields have doubled since 1961 (Runge et al. 2003). It appears that additional genetic increases are still possible because adequate genetic variability exists. Even in developed countries, average yields are far below record or theoretical yields due to the effects of drought and weather-related problems, biological constraints, or soil-related problems (Boyer 1982). Significant differences in yields among countries with similar yield potential emphasize the opportunity for the broader use of conventional technologies. Average U.S. maize yields of 8,900 kilogram/hectare (kg/ha) compare with those of Kenya at 1,500 kg/ha. Rice yields in China of 6,100 kg/ha compare with 2,500 kg/ha in Thailand. Average wheat yields in France of 6,200 kg/ha are double those of Russia (FAO 2004a). While the differences in these comparisons are accounted for by a number of factors, much of the difference is a result of differences in the use of conventional technologies.

Government policies have distorted the economic feasibility of technology adoption in some cases. In the EU and the United States, subsidized water, subsidized energy, and large crop subsidies for farmers have encouraged the widespread adoption of high-technology farming methods.

Among the many conventional technologies currently in use, six general categories offer particular promise for increased productivity and sustainability if their use were expanded.

1. Plant breeding and other conventional genetic technologies, which have been responsible for significant increases in crop yield, have not been adopted everywhere. Yield improvement in most crops is still possible through traditional breeding programs in varietal, non-hybrid crops. Very little effort has been devoted to crops that are important to smallholder farmers compared to the large, private-industry effort for commercial, developed-world crops. In spite of increases in overall global productivity, crops of importance in developing countries such as cassava, sweet potato, sorghum, pearl millet, cowpeas, and so on, have not benefited from plant-breeding-based improvements in productivity. As was the case with the "green revolution," genetic improvement is often expressed most completely when combined with irrigation and fertilizer use.

2. Hybrid crops are considerably more productive than varieties, but have only been adopted in some areas. For reasons related to the amount of hybrid vigor and practicality of seed production, not all crops can be hybridized, but many can. Hybrid maize is grown on less than 50 percent of the world's maize acreage, and hybrid rice is only 50 percent of China's rice acreage in 2001 and much less in the rest of the world. Virtually no hybrid wheat is currently

grown, although hybridization of wheat is technically possible and practical methods of hybrid wheat-seed production are likely in the next 10 years. Considerable infrastructure is required to produce hybrid seed and provide it to farmers every year, since farmers experience a drop in productivity if they use the grain from their hybrid crop for seed the next year. The cost of developing this infrastructure may be one of the reasons for the lag in the adoption of hybrid technology. Genetic systems, like apomixis, which fix hybridity and reduce infrastructure associated with hybrid crops, have received only a small amount of research attention, but hold great promise for the developing world (Grossniklaus et al. 1998; Jefferson 2001). Using apomixis technology, hybrid seed no longer needs to be produced by crossing inbred lines. With apomictic seed, farmers can realize hybrid performance but still save seed for the next generation since the hybrid vigor is made permanent through a genetic mechanism.

3. Pest control in crops can be addressed by a number of existing technologies. Pesticides and herbicides have increased crop productivity in many parts of the world. Unfortunately, serious pollution has resulted. In some cases, problems arose from the toxic nature of the chemical itself and in other cases from misuse or overuse. Replacement of older, more dangerous chemicals with newer, less-toxic chemicals would have an overall benefit. At the same time, pesticide use could be minimized by adoption of integrated pest-management strategies. Use of biological sources of resistance, both conventional and transgenic, can also replace more hazardous chemical approaches and can be increased in world agriculture.

4. Fertilizers are available and widely used, resulting in favorable return on their cost. Fertilizer use in the developed world decreased in the late 1980s and early 1990s by 30 percent. Fertilizer use in sub-Saharan Africa is less than 10 kg/ha of cultivated land (Harrison and Pearce 2000). Fertilizer use increased in Asia during the "green revolution" and in some cases exceeds use in the developed world. Lack of infrastructure for both fertilizer production and distribution, along with high cost, inhibit its use in many areas where it could increase crop productivity. Clearly, such use can be excessive. For example, fertilizer use in the United States and other developed countries has resulted in degradation of water quality. Among the most egregious cases is the formation of the "dead zone" in the Gulf of Mexico, a 22,000 square kilometer area where nutrient runoff from the Midwest has caused eutrophication and resulted in the death of fish and other marine life (Holden 2002). Cropping systems, including reduced tillage, intercropping, rotations, and manure-based fertilizers, are beneficial in many locations, but are still underexploited.

5. Freshwater may well be the most limiting factor for global agriculture (Smil 2000). Irrigation technologies for water distribution and conservation are known. Techniques and hardware for improving irrigation efficiency include direct seeding, drip irrigation, precision sprinklers, micro sprinklers, and irrigation scheduling/timing (Gleick 2003). Only 11 percent of Africa's

farmland is irrigated (Runge et al. 2003). Competition between urban and rural water users is accelerating in many arid regions of the world. Despite availability issues and problems with salinity and sodicity, vast increases in productivity through irrigation are possible.

6. Better utilization of post-harvest storage technologies could increase developing-world food availability. For example, an estimated 40 percent of the developing world's harvest is lost after removal from the field (Pimentel 1976). Rats, insects, and fungi all share the crop, inadequate storage being a problem. Technologies exist to control all of these losses and are particularly valuable in the tropics where the losses are the greatest.

Recognition of the potential of these underused technologies is not meant as an unqualified endorsement of their use throughout the world. The adoption of some of the conventional technologies examined above can have detrimental effects on the environment when used in excess or in inappropriate situations. Movement of developed-world technologies around the world is not always warranted, but in some cases current problems experienced in the developing world could be reduced with the appropriate application of these technologies.

New Technologies and Traits

The focus of this paper is biotechnologies as they affect crop agriculture. Advances in plant molecular biology have given rise to agricultural biotechnology. New discoveries have allowed scientists to better understand the genetics of crop plants. These advances have also enabled new and efficient technological capabilities including gene isolation, identification of gene function, and the movement of genes across species boundaries. A transgenic crop is defined as one that contains a gene or genes from a different species or synthetic genes introduced through the tools of genetic engineering.

The application of these scientific advances will result in traits valuable to the farmer, the consumer, and the pharmaceutical and other industries. Advances in science will continue to make the manipulation of plant genetics easier and more cost effective. The introduction of genes for new traits in crops and two of the most important enabling technologies for transformation/gene insertion and genomics are discussed below.

Traits for the Farmer

Genes coding for proteins that confer a broad array of traits that have not existed in crop plants have been identified and isolated from non-crop plant species and bacteria. The most important of this first generation of traits include insect resistance, virus-disease resistance, and herbicide resistance. A more complete list of current and future traits is shown in table 1. Farmers in some countries have rapidly embraced seed varieties and hybrids containing the first generation of these transgenic traits. Transgenic crops with improved insect and herbicide resistance have been commercialized in soybeans, maize, cotton, and canola.

Eighteen countries, led by Argentina, Brazil, Canada, China, and the United States, grew 67.7 million hectares of transgenic crops in 2003 (James 2003). Farmers have benefited economically and have realized increased flexibility in their farming operations through the use of these traits (Marra 2001). To date, these traits have not resulted in detrimental human health effects, which were feared. Growth in the area devoted to these crops is likely to continue and spread to other countries in the Americas, Asia, and southern Africa.

Single gene traits, such as insect resistance, virus resistance, herbicide resistance, and feed quality have been relatively easy to develop. These first-generation traits were pursued, in part, because of their technical feasibility. They have primary value to farmers in replacing another input like a chemical insecticide or by increasing yield by preventing unrecognized yield loss caused by insects or by allowing the use of a less-expensive, more easily managed weed-control program. Traits controlled by multiple genes, such as disease resistance, nitrogen use efficiency, drought tolerance, salinity tolerance, and increased yield are more technically challenging. Even so, it is likely these traits will be available to farmers in some form before 2020.

Transgenic crops that solve insect, disease, and agronomic problems are increasingly in demand in the developing world where these problems tend to be more serious because of the tropical environment of many developing countries and because other options like pesticides and irrigation are less available.

Traits for the Consumer

Traits of greater interest to consumers rather than farmers are often cited as having the potential to increase the public acceptance of transgenic crops, offsetting concerns about the technology. Traits offering benefit or appealing directly to consumers have not reached the market, although several such traits are now on the path to commercialization (BASF, Bayer Cropscience, Dow Agrosciences, DuPont, Monsanto, and Syngenta Web sites 2004). Food companies tend to be conservative and market-research driven. They will probably move cautiously to adopt transgenic crops with consumer benefits. Their experience to date has been somewhat negative. They have had to deal with the controversy over genetically modified plants, but they have not received the financial benefits, which have mostly gone to farmers and seed companies. It is unlikely that a trait with widespread consumer appeal in the developed world will reach the market before 2015.

Traits offering improved nutrition (increased vitamin A, iron) and other potential benefits for people in developing countries are a different story. These traits may be met with enthusiasm and contribute to solving malnutrition (FAOa 2004; Interacademy Council 2004). Limited scientific capacity, lack of paths to commercialization (business infrastructure), and the lack of a regulatory infrastructure in many developing countries will impede development and dispersal of these traits. The production of vaccines in crops like banana for consumption by humans in the developing world has received considerable research attention in the public sector and is much less controversial. Genes

Table 1. Current and Future Genetic Coding Traits

Traits for Farmers	
Insect resistance	Herbicide resistance
Virus-disease resistance	Fertilizer-use efficiency
Fungal-disease resistance	Drought tolerance
Mycotoxin degradation	Adaptation to acid soils
Salinity tolerance	Parasitic-plant (Striga) resistance
Heat and cold stress tolerance	
Hybridization methods (apomixis)	
Post-harvest loss prevention	
Nematode resistance	
Animal feed quality	
Phosphorous availability (phytase)	
Traits for Consumers	
Developed World	***Developing World***
Modified fatty acids	Enhanced vitamin A
Increased antioxidants	Increased iron
Decreased allergens	Disease-resistant banana
Expanded shelf life	Plant-produced human vaccines
Increased omega fatty acids	Plant-produced animal vaccines
Decaffeinated coffee	
Increased or modified protein	
Traits for Industry	
Pharmaceutical production in plants	
Specialty chemical production in plants	
Biodegradable plastics	
Industrial enzymes	
Pigments	
Industrial oils	
Castor oil without ricin	
Plants that absorb heavy metals (remediation)	
Increased pulp production in trees	
Low lignin production in trees	
High-amylase corn for ethanol producers	

coding for bacterial proteins that elicit antibody formation in humans but do not cause disease are being developed for Vibrio cholerae and enterotoxigenic Escherichia coli in an attempt to create an edible vaccine in bananas (Walmsley and Arntzen 2000).

Traits for Industry

In addition to traits of value to farmers or with appeal to consumers, biotechnology enables the production of additional products in plants. Plants can be transformed with genes that will enable them to produce both pharmaceutical and industrial compounds. Pharmaceutical production in plants may be more cost effective than the production of drugs through current methods like fermentation. Several companies have businesses aimed at pharmaceutical production in plants, including Prodigene and Dow (Giddings et al. 2000). The increased number of protein drugs coming from medical biotechnology increases the value of pharmaceutical production in plants since fermentation capacity may become a limiting factor, and plants have advantages over bacteria for protein production. Further experience is needed to determine whether the economics are attractive to the pharmaceutical companies and whether adequate methods exist to ensure segregation of the plants with these traits from plants in the food supply (Cassidy and Powell 2002).

Production of industrial products by plants such as rubber, wood, cotton, and linseed oil is familiar. Transgenic plants may be modified to produce other industrial products or provide intermediates that replace petroleum-derived compounds (see table 1). Biodegradable plastics have been produced in plants (Somerville and Bonetta 2001). The relatively low price of petroleum and considerable investment in the infrastructure to manufacture industrial feedstocks have represented barriers to the introduction of a new method of production. While many interesting compounds can be produced in crops, they are generally not cost competitive with existing methods of manufacturing. It is possible that in the longer term these approaches will represent an alternative to increasingly scarce and expensive petroleum. Significant changes should not be expected to occur by 2020, but several companies continue to focus on this opportunity.

The tendency to judge the value of transgenic crops on the basis of the few traits that have been developed to date inappropriately limits the discussion and undervalues the potential importance of the technology. The technical feasibility and eventual market value of various traits is difficult to predict. It is worth noting, however, that the technology remains in its infancy. A number of forces will continue to act to improve the level of scientific understanding and possible applications for the technology in the future.

New Technology: Gene Insertion into Crop Plants

Techniques for inserting genes into crop species (transformation) and achieving stable gene expression without negative effects on the plant are an essential first step for modifying crops. Existing methods for inserting genes into crops are labor intensive and limited to a few crop species and a small number of cultivars within those crops. Transformation is an active area of research, particularly in the private sector. Between now and 2020 we can expect the development of non-crop-specific, efficient, site-specific techniques for plant transformation. The resulting transgenic plants will speed hybrid and variety development as the effects of random gene insertions will no longer need to be characterized and evaluated in the field. These new transformation techniques may also simplify regulatory review since the gene insertions will occur in the exact same location in the genome every time. Unpredictable insertion events will not occur and thus transgenic plants will require less molecular characterization for regulatory purposes.

Scientists in developing countries will be able to apply improved gene-insertion techniques to crops of interest to smallholder farmers and develop traits that affect human nutrition. Effective transformation systems for producing transgenic versions of crops grown in developing countries, such as cassava, yams, and cowpeas, do not exist. This gap in scientific capability prevents the application of genetic engineering to these crops regardless of the potential value of the desired trait. Companies are currently conducting most of the research on transformation. Therefore, progress in this area may raise questions related to corporate control of intellectual property. Patents on first-generation technology for transformation will expire before 2020, thus making these techniques available to developing countries. However, new and more powerful methods may not be readily available unless companies provide access to their intellectual property in crops that have little commercial value for them.

New Technology: Genomics

In the 25 years since the origins of applied plant molecular biology, the identification of useful genes and subsequent efforts to move these genes into crops has been slow, labor intensive, and inexact. That situation has changed in the last 10 years and will continue to change for several reasons. A new approach to the study of genetics, called genomics, a term that refers to study of the entire complement of an organism's DNA or all of its genes, has increased the tools for understanding the basis of inheritance. The development of high-throughput systems for sequencing DNA, understanding gene expression, and profiling the protein makeup of organisms has driven a revolution in the understanding of biology. Combined with algorithms, software, and computer systems to analyze the sequence information, our understanding of how plants and animals function has exploded.

Interest in genomic information to understand the medical biology of humans and human diseases has driven high levels of public and private spending that has resulted in spectacular progress. Clearly, this kind of spending did not occur with agricultural biotechnology in mind. However, since DNA and other molecular components are common across all species, the technology is readily applicable to plants, fish, or farm animals. Spending on the development of important tools or "omics" technologies for delivering information has resulted in an impressive array of tools. Among these tools are DNA sequencers, chip-based systems for measuring gene expression (which genes are making proteins at different times and in different parts of the organism) for thousands of genes simultaneously. Other instruments measure the levels of all an organism's proteins (proteomics). Mass-spectroscopy-based approaches measure the complete complement of metabolites in an organism (metabolomics) (Hirai et al. 2004). Perhaps most importantly, in addition to massive levels of spending for tool development, considerable private and public funding to apply these tools directly to crops has occurred in the last 10 years.

Concentrated efforts to sequence the entire genome or at least all of the expressed genes of a number of crop species have resulted in significant basic information about plant biology. Complete genome sequences are now available for *Arabidopsis thaliana* (an experimental plant species used by scientists to understand plant biology) and rice. Most of the genes coding for proteins from maize, wheat, and soybeans have been sequenced as well. Genes can be identified and isolated using tools from the various "omics" technologies.

Pieces of DNA that control and coordinate these genes are also being discovered. These discoveries allow scientists to determine the role a gene product or protein plays in the plant. Portions of the DNA called regulatory sequences modulate control of genes so that they are turned on and off at the appropriate time and level. While plants have tens of thousands of genes, more than 90 percent of the genes code for proteins that have similar functions across species. This similarity makes it possible to use species with relatively simple genomes, like bacteria, *Arabidopsis,* and rice, as well as the increasing number of more complex but fully sequenced organisms like humans and yeast, to provide information that is readily applicable to plant species like maize or wheat.

Significant private and public funding of plant-genomics research will continue to drive the discovery of useful genes. The National Plant Genome Initiative is aligning its goals with the expectation that the National Human Genome Research Institute will have developed capabilities for sequencing an organism's genome for $1,000 by 2013 (National Plant Genome Initiative 2003). Concomitant progress in bioinformatics to analyze and compare genomes will also be driven by funding for human applications.

This investment and the resulting scientific progress indicate that by 2020 the following situation can be expected:

- The complete genome sequence will be known for all crop species.
- Significant sequencing of non-crop plant species and microbial species will occur.
- Sequence information along with functional genomics will result in knowledge of the function of over 95 percent of the genes of crop plants and genes from other sources.

- The function of genes for entire metabolic pathways for valuable secondary metabolites (taxol, etc.) will be known, enabling the movement of suites of genes.
- Sequence information will be useful to make diagnostics to identify diseases and other problems and molecular markers to improve the efficiency of plant-breeding programs.
- The understanding of inheritance and variation among individuals will allow scientists to predict the outcomes of crosses of lines within plant-improvement programs.
- The conservation of genetic information will make these techniques applicable across all types of crops and traits.

The application of these technologies will provide genes to impact a broad number of traits, including those of interest to farmers, consumers, and industrial users of the technology. New scientific discoveries will increase our knowledge of genes and their functions. This knowledge will be used in the development of new cultivars that will improve agricultural productivity.

Capacity: A Profile of Global Research and Innovation

The global scientific capacity now exists to take advantage of current and future scientific discoveries. Substantial public and private research investment in plant biology, and leverage of basic molecular-biology funding by the National Institutes of Health (NIH), pharmaceutical companies, and biotechnology suppliers have resulted in the current crop-biotechnology capacity. Research spending by the private sector on crop-biotechnology and genetic-crop improvement is approximately $1.2 billion per year (various company Web sites). The development and introduction of the first transgenic products resulted in a restructuring of the seed and agrichemical industries, which continues today. Consolidation of the companies in these areas occurred through mergers of the agrichemical companies and acquisition of the seed and biotechnology companies by agrichemical companies (Thayer 2001). In 2004, there are six major companies: BASF, Bayer, Dow Agrosciences, DuPont, Monsanto, and Syngenta. As recently as 1992 there were 16 companies: American Cyanamid, BASF, Bayer, Ciba-Giegy, Dow Chemical, DuPont, Hoechst, ISK Biosciences, Merck, Monsanto, Rhone Poulenc, Rohm & Haas, Sandoz, Schering, Shell, and Zeneca.

Several waves of small venture-backed companies in the United States and Europe have occurred since the early 1980s. The original wave of biotechnology companies, focused on trait discovery, were acquired over the course of the 1980s by large companies interested in building seed businesses. Another wave of "genomics companies" still exists but has been restructured. Companies that use functional genomics and other related techniques to discover genes, companies that develop pharmaceutical production in plants, and basic-technology companies are operating on venture funding while they wait for a window in the initial public offering (IPO) market or acquisition by the large players. Venture money is still available, but concerns about the acceptance of the technology and the unfavorable IPO market have limited enthusiasm. Nevertheless, companies continue to be brought forward.

Scientific research and product development, by both the large and small companies, has been concentrated in the United States, Canada, and to a lesser extent Europe and Australia. It is likely that private research spending will continue to be concentrated on a few crops (maize, cotton, canola, and soybean) for the developed-world market. Slow acceptance of biotechnology in Europe may gradually open additional markets, but not before 2010. Continued investment by the multinational corporations should make the technology available in areas with favorable economic incentives. Even so, there are too many multinational companies and too much research spending for the size of the seed market, and two or three of the current players could exit the business. Venture capital will remain available, although at a lower level and may drive additional waves of small-company development based on new technological developments.

Public spending on plant molecular biology in the United States has been at approximately $300 million per year in recent years (personal communication, Brian Hyps, American Society of Plant Biologists). Funding growth in the United States for the National Science Foundation, U.S. Department of Agriculture (USDA), and the NIH will decrease as attempts to fund homeland security and weapons development and reduce the federal budget deficit continue (Teich and Koizumi 2004). Small but meaningful public research spending in the European Union and Australia adds to the effort. Most publicly funded research in plant molecular biology is conducted using the experimental plant species, *Arabidopsis thaliana*. With a small, sequenced genome and a short life span of weeks, the use of this model species has accelerated progress in the last 10 years. However, discoveries made in *Arabidopsis* must then be applied to crop species with many more genes.

Research Capacity in the Developing World

Many of the applications of agricultural biotechnology are scale neutral, when farm size is considered, so adoption is not limited to the large commercial farms. In fact, it is possible that adoption of certain transgenic products might allow developing-world farmers to bypass certain problematic technologies (insecticides, fungicides) or technologies requiring large infrastructures (hybrid crops). A look at transitional economies and those of Africa reveals some of these issues.

Capacity issues concerning China and India deserve examination. Both of these countries are strengthening their positions in agricultural biotechnology (Huang et al. 2002a; Huang et al. 2002b). In many cases, agriculture in China is already quite advanced. Nearly 100 percent of the maize in China and 50 percent of the rice is grown from hybrid seeds. China is also in the process of privatizing the national seed industry. China's desire to feed its own population is driving investment in agricultural technology. China is currently building a significant internal-crop biotechnology capability. The Beijing Genomics Institute was the first group to sequence the genome of indica rice (Yu et al. 2002). This group looks and acts much like a U.S.-based biotechnology company. The Chinese are developing their own version of many transgenic products and have been particularly effective in solving problems in rice.

In recent visits to seven Chinese research institutes in Beijing and Shanghai, this author learned that six were in the process of building major new facilities. In many of these institutes, large research groups are headed by Chinese-born, Western-educated scientists who have been in the United States or Europe for a considerable part of their careers. Labs are generally well staffed but are not presently equipped at the level of those in the United States and Europe. There is concern that current U.S. visa requirements have reduced the number of Chinese students coming to the United States to study. In the past, large numbers of Chinese students received their graduate training in the United States and then stayed for university and industry jobs. A reduction in students from outside the United States, combined with the falling numbers of U.S.-born students interested in science, could have serious implications for a loss of U.S. technical leadership (Mervis 2004).

India also has a nascent but potentially competitive scientific infrastructure in the agricultural-biotechnology area. Around U.S.$2 billion is invested in the Indian biotechnology sector (all types), and approximately 25,000 people are employed. The government hopes that by 2010 more than 1 million jobs and a U.S.$10-billion investment will exist (Madhavan 2004). India has a meaningful agro-industrial capability as a result of being the primary beneficiary of the green revolution. Adoption of Bt cotton is benefiting Indian farmers (Qaim and Zilberman 2003). In addition, low costs and a well-trained scientific population have made India a favorite location for offshore research efforts by multinational corporations.

The decision that countries make about growing increased hectares of transgenic crops is driven at least in part by whether they export grain to Europe and Japan. Argentina, Brazil, Canada, and the United States, the countries that have embraced biotechnology, are all grain exporters. Since many of theses exports are to Europe or to Japan, the exporters are all constrained by the acceptance of transgenics by the importing countries. Cases at the World Trade Organization (WTO) arguing unfair trade practices have been filed by Argentina, Canada, and the United States against the European Union for its unwillingness to accept transgenic products, but these cases have not yet been resolved. Non-transgenic sources of some crops, particularly soybeans, could become more difficult for the Europeans and Japanese to obtain.

China and India have allowed farmers to grow a limited number of transgenic crops, particularly cotton, and are conducting field tests, but they are still debating the merits of growing increased areas of transgenic crops. Their decisions about how widely to use the new technology will be driven by their attitudes toward multinational corporations, internal technological capability, food needs, and their analysis of the risks of transgenic crops. Since neither China nor India exports significant amounts of their production to the EU, they will not be required to consider the opinions of the Europeans. This situation offers them exceptional freedom and autonomy to make their own decisions.

In the years between now and 2020, China and India will strengthen their positions in agricultural biotechnology. In the longer term, China and India could become the center of scientific capability in agricultural biotechnology in the world.

Africa is in the most tenuous situation regarding the development of transgenic crops. These countries have the highest rates of malnutrition and need to increase their capacity for food production. At the same time, they have hopes of exporting agricultural products to Europe and would like to avoid the risk of losing this potential market. Europe's position would appear to be contrary to the needs of Africa. For example, international agencies like the Consultative Group on International Agriculture (CGIAR) have reduced funding for agricultural improvement and directed monies away from biotechnology approaches for crop improvement in an attempt to appease European funders (Lele 2003).

The increase in private research investment to develop transgenic crops has been targeted at the large-area, developed-world crops like maize, soybeans, and cotton, particularly in North and South America. Scientific research to address the opportunities in the developing world is quite modest. There are cases where genes that are useful to the developed world can be of interest in the developing world, but work on crops—rice, cassava, and millet—and traits of value to farmers in developing countries is largely missing (Jefferson 2001). Genes for insect resistance, like the *Bacillus thuringiensis* toxin genes, can be of value to farmers throughout the world, and are readily transferable for use in different crops. Many traits, such as human nutritional quality, virus resistance, disease resistance, apomixis, and drought tolerance are of far more interest to farmers in developing countries. Multinational corporations invest in the development of traits for the developed world where commercial farmers will provide the companies with a reasonable return on their investment. Companies will not, given their fiduciary responsibilities to shareholders, invest in research and development on traits that are primarily of interest to developing-world farmers unless they can identify a reasonable business model. The question then becomes, who will develop products for the developing world?

A number of reports (FAO 2004b; Interacademy Council 2004) promote the potential value of transgenic crops to satisfy hunger in the developing world. If the value is to be realized, the need for research to drive the development of these seed varieties must be addressed. In addition, the need to develop production and distribution infrastructure in the developing world is critical. Though it is unrealistic to expect every developing country to create an internal biotechnology capability, the transgenic technology will have to come from somewhere. The FAO *State of Food and Agriculture* (2004b) report urges developing countries to focus their attention on traditional breeding and regulatory systems instead of investing in the research capacity needed to utilize biotechnology. It begs the question of who will fund research to create transgenic crops and traits of value in the developing world.

As pointed out earlier, China, Brazil, and India possess the capability to produce transgenic products of their own and are currently pursuing this approach to crop improvement. Other countries, particularly those in Africa, cannot afford internal programs and are dependent on outside sources of technology. In a recent study of the scientific importance of the 31 countries that produce 97.5 percent of the world's scientific output (in terms of cited papers), only one country in Africa—South Africa—is on the list, and it is number 29. China and India are numbers 19 and 22, respectively (King 2004). The state of basic discovery research is a good indicator of research capacity, since research in plant molecular biology is often readily transferable from basic discovery to applied purposes.

Rapid progress in basic research carried out in university laboratories in both the developed and developing world is applicable to building developing-world research capacity. Even so, there is considerable work involved in making the discovery of an interesting gene into a seed product that a farmer can use. The gene must be put into a crop where it must be expressed in a way that results in a change in the trait. It must also be incorporated into a variety of the crop that is adapted to the local growing environment in terms of length of growing season, disease and insect resistance, photoperiod responsiveness, and other non-transgenic traits. Most developing countries are a long way from reaching the capacity needed to produce transgenic crops for their areas of adaptation (McElroy 1999).

Several organizations are providing funding for agricultural research in developing countries, but the efforts to develop capacity in biotechnology remain underfunded. The CGIAR had a major role in the green revolution. Today, it is still the major international research organization involved in improving agriculture in the developing world through research. CGIAR spends about U.S.$25 million per year on biotechnology (Lele 2003). This spending is very diffuse as it is spread out over a wide range of crops and research centers. Decreases in CGIAR's budgets and increases in the scope of projects to include social and environmental efforts, along with the resistance of some European contributors to fund biotechnology, have resulted in a weak effort. The Rockefeller Foundation distributed more than $100 million for research on the basic biology of rice (O'Toole et al. 2001). More recently the Gates Foundation has indicated interest in funding research related to the nutritional value of crops (Varmus et al. 2003). The U.S. Agency for International Development (USAID) has also funded research in biotechnology for the developing world.

In the short term, the discussion of the potential value of biotechnology for solving the hunger problem in the developing world may be moot. Even if everyone agreed that the technology is of value, which is certainly not the case, progress is unlikely unless a number of organizations, mostly public organizations, are willing to make a commitment to address the problem. The concern over access to intellectual property has been overemphasized, perhaps in part to reflect an assumption that traits of value to the developing world are the same as those of value to the private companies involved in biotechnology research. A lack of investment has left developing countries without the research capacity or the production and distribution infrastructure needed to take advantage of the new biotechnologies.

Challenges for Global Research and Innovation

Scientific progress will drive understanding of genetics and discovery of genes and traits that have use in agriculture, but there is a very real question as to the extent to which this basic understanding of crop plants will be applied. Many factors will impact the potential adoption of transgenic crops and the eventual impact on the relationships among countries. Issues related to government regulation of the technology, intellectual property, scientific and agricultural capacity and infrastructure, global agricultural policy, public acceptance, and environmental impact will all affect the extent of adoption and the resulting impact on the relations of regions and countries.

Regulation of Biotechnology

Governments have arrived at very different methods of regulating the products of agricultural biotechnology in order to ensure food and environmental safety (Pew Initiative on Food and Biotechnology 2004). There is general agreement that regulatory review of transgenic products to evaluate risks and establish public confidence is a necessary requirement for commercialization. The main risk factors—toxicity, allergenicity, effects on non-target insects, gene flow to surrounding crops and wild relatives, and development of resistance to insecticidal proteins or herbicides—are well known (Pew Initiative on Food and Biotechnology 2004). Yet, despite this knowledge and agreement, divergent views exist on the extent of the risks and the regulations needed.

The current regulatory system in the United States is based on a model from the agricultural-chemical industry and requires the involvement of considerable supporting paperwork and significant cost. Public confidence is enhanced by a transparent regulatory system that allows examination of data submitted in support of a new product. However, company needs for confidentiality of critical business information have raised suspicion among groups that are opposed to the technology and feel they lack full access to critical data.

The EU and Japanese regulatory systems have been much more cautious than the U.S. system and to date have all but prevented the introduction of transgenic products to their farmers. Many transitional-economy countries (China, Brazil, Argentina, India) have taken an intermediate stand and have allowed the introduction of transgenic crops. They have moved more slowly and in a more limited way than the United States (James 2003). Many African countries lack the money and scientific expertise to establish effective regulatory systems and have, as might be expected, moved slowly to establish these systems. It is essential that these countries develop the internal capabilities to make decisions for themselves if they hope to participate in the world agricultural market (Pinstrup-Andersen and Schioler 2001).

One of the most controversial issues related to the regulation of biotechnology is the labeling of foods made from transgenic crops. In general terms, labeling has been required in Europe but not in the United States. Arguments over the value and adequacy of this information and whether the impact of the information is positive or negative continue. The large food companies oppose labeling. It is often thought that such a requirement in the United States could result in the food companies' withdrawal from producing foods made from transgenic crops.

Currently, there is a much higher standard for regulatory review of transgenic crops than for crops developed in other ways (mutation breeding, traditional breeding, backcrossing, etc.). The high cost of regulatory approval, which ranges from U.S.$10–$25 million for a new transgenic trait, limits attempts to develop transgenic products to those crops and traits with the highest potential economic return. Small crops or traits that may have positive environmental consequences but no profitability advantage for farmers do not make the economic hurdle. An example of a trait that has not met the economic hurdle is low-phytate grain crops, which would reduce phosphorous in manure and eventually in the water supply.

In order for the developing world to fully utilize the science now available, it will be necessary to minimize the cost of the regulatory regime while still ensuring food and environmental safety. Crops and traits of interest to smallholder farmers in developing countries are unlikely to find their way through a high-cost regulatory system. Concentration of ownership and development of transgenic crops will be limited to large multinational corporations as long as the regulatory costs remain high. USDA, universities, the Danforth Center, and other public research groups have a number of traits that could be commercialized or made available to farmers, but these traits may not advance past the current regulatory hurdles.

Food-safety data, toxicity, and allergenicity tend to be applicable across human populations and may be transferable across different crops with the same traits. Sharing safety data across country boundaries would simplify and lessen the cost of current regulatory systems. Data on environmental effects tend to be more specific to a given location and ecosystem. The presence of different insects or weeds, different types of agronomic practices, and the presence or absence of a crop's wild relatives all require location-specific regulatory evaluations and must therefore be more country or region specific. In addition, environmental effects may be more likely to be expressed over a longer period of time, requiring more long-term monitoring (National Research Council [NRC] 2002b).

Regulation of crops or plant species with added pharmaceutical or industrial traits requires a more comprehensive and wide-ranging regulatory review than other transgenic crops. Potential harm resulting from certain pharmaceuticals or industrial products entering the food supply requires methods of segregating these crops from those used for food. Segregating mechanisms that allow complete containment of the transgenic crops are difficult to implement with certainty. It has been argued that the difficulty of complete crop segregation requires the use of nonedible crops or plants for these purposes (Cassidy and Powell 2002).

Global agreement on the main factors to be regulated and cross-border sharing of data would greatly increase the likelihood of evaluating the risk of a transgenic crop in a cost-effective way. Since many of the benefits of transgenic crops may accrue to countries that can least afford expensive regulation, these countries' eventual access to the technology is dependent on the development of these systems.

Intellectual Property Protection

The "green revolution" of the 1960s was accomplished with publicly available technology. Two international agricultural centers, the Center for the Improvement of Maize and Wheat and the International Rice Research Institute, were publicly funded and provided new cultivars of rice and other grains (Evenson and Gollin 2003). The adoption of these new cultivars, combined with increased irrigation and fertilizer use, had a dramatic effect on rice and wheat yields in Asia and to some extent in Latin America. Since that time, the landscape for the protection of intellectual property has changed.

U.S. Supreme Court decisions in the early 1980s first provided for patents on living organisms (Diamond v. Chakrabarty) and then created patent protection for genes, hybrids, and methods for improving crops (*Ex parte* Hibberd). As a result, private research investment in genetic improvement of crops by biotechnology and traditional breeding rose to exceed public investment (FAO 2004b). Large investments in biotechnology and plant breeding by the multinational companies made sense once their return on investment could be protected by the temporary (20 years from the date of filing) monopoly provided by patents. Farmers could no longer save the seed of patented material, and price premiums were justified from protected cultivars. It is no surprise that real progress in biotechnology has only occurred in those traits, crops, and countries where significant markets are coupled with viable intellectual property protection.

Controversy over the ownership of genetic resources and the concentration of power over agriculture in the hands of large corporations followed the court decisions to allow patent protection for genes, hybrids, and methods for improving crops. Opponents of the technology, while concerned about food and environmental safety, have been more likely to base their concerns on the control of the technology and "corporate grab for a public resource" (Cayford 2004). Concerns about high inputs and industrial-style agriculture have generally led opponents to oppose the technology rather than attack the underpinnings of the legal system that has made it possible.

Several other specifics related to patenting have escalated concerns. In the early years of the development of biotechnology, the U.S. Patent Office was willing to grant very broad patents. Some of these patents covered all crops or all dicots rather than a single species, even though the invention had not been enabled broadly. In many cases, these broad patents inhibited public researchers from working in similar areas or using what they believed to be patented technology. At the same time that complaints were being aired about corporate control of intellectual property, university-based researchers, motivated by the Bayh-Dole Bill, began patenting their discoveries as well (Nottenburg et al. 2001). University patent offices would then seek to license the rights to the patents (often exclusively) to the highest bidder (often a large corporation) in return for a royalty or other financial reward. A corporation could now obtain rights to research results that might once have remained in the public domain. Attempts to create products for use by farmers in the developing world have been frustrated by concerns about freedom to operate. There are several groups that are attempting to make intellectual property available for use in developing countries, but little research is actually done with the intent of increasing public research capacity.

Legislation to overturn patent protection on plants and genes is unlikely. In spite of this, there are several reasons that the technology necessary to enable transgenic crops will find its way into the public sector.

- First, patent rights only extend for 20 years from the date of filing. In the case of many first-generation biotechnology patents, rights will expire before 2020 and the technology will become available.
- Second, universities that have aggressively patented and licensed technology might be convinced to forgo exclusive licenses and to hold back rights for technology of interest in small crops or in developing countries. For example, recently, the Rockefeller Foundation and 10 U.S. research institutes and universities reached an agreement on a mechanism to forgo exclusive licensing and withhold rights of patented inventions for developing countries (Atkinson et al. 2003).
- Third, patents are a country's right and apply only within the country in which the patent has been filed. Since filing fees can be significant, companies generally file only in countries in which they will likely have a business. Patents are not often filed in developing countries since companies do not intend to do business there and the enforcement of the country's intellectual property laws may be questionable. If one assumes that research capacity exists, research based on inventions patented in developed countries could be carried out in the developing world for use there.
- Fourth, it is not clear what path China, Brazil, Argentina, and India will take regarding intellectual property rights. In some cases in the past, these countries have taken a fairly nonchalant approach to intellectual property protection. In the case of Argentina, Monsanto stopped selling transgenic soybeans because of the lack of adequate intellectual property protection.
- Fifth, large multinational corporations like DuPont, Monsanto, and Syngenta have been willing to make technologies available for use in developing countries. Making these technologies available has been in the companies' interest since it results in favorable public relations. These actions have also been seen as a way to influence Europe by broadening the adoption of biotechnology in developing nations where people are hungry. The companies

are also motivated by the desire to do good in those areas where they cannot yet do business in order to create goodwill for possible future business opportunities. The largest constraint for companies donating technology to the developing world is the concern over liability. For example, a product developed by a public organization and containing a company's technology might harm someone. In this case, the companies need to be able to avoid liability where they no longer control the use of their technology. This problem has not yet been solved. A well-respected foundation or nongovernmental organization such as the Bill and Melinda Gates Foundation or the Rockefeller Foundation might be able to intervene to find a solution.

Intellectual property issues will slow the implementation of agricultural biotechnology, but it is unlikely that these issues will be as large a factor as regulatory concerns and the lack of scientific capacity addressing the needs of the developing world. Intellectual property issues can be addressed in a number of ways. With more in-depth knowledge of how patents work, public-sector research organizations will better understand their options and may make different decisions. Wealthier countries with strong intellectual property laws will continue to see investment in new technology. Concerns over corporate control of agriculture and food will persist.

Agricultural Policy Relationships between Countries

Developed countries have supported their farmers to maintain national food security, address the needs of rural political constituents, and maintain favorable trade balances. The United States and European Union subsidize their farmers on the order of $70 billion per year (Thurow and Kilman 2003). In spite of consolidation of farms and a steady decline in the number of farmers and farm workers, subsidy payments continue to be a large percentage of income for farmers in the developed world. Large-scale farmers in the United States have adopted mechanical, genetic, chemical, and computer technologies to increase productivity and cut costs. While the breadth of the United States' geography and farm size encourages these techniques, farmers within the European Union have continued to farm smaller acreages and to use more traditional farming methods. Maintaining the appearance of the landscape is also more important to European farmers. It is worth noting that a small but rapidly growing organic sector has developed in response to consumer demand. Organic farms vary in size and use traditional farming methods. The USDA "certified organic" definition does not allow the use of transgenic crops, although the logic for this decision may be more political than technical.

Agricultural subsidies for certain crops in the United States, Europe, and Japan have resulted in the charge that the low prices caused by these payments deny developing countries the chance to create their own agricultural exports, thus excluding them from one of the few areas in which it might be possible for them to compete with developed countries. The difficulties encountered in the Doha Round of WTO trade talks in Cancun, Mexico, reflect the differing points of view of the developed and developing countries in this debate. In a few cases, acreage of some crops has shifted to developing countries. Soybean acreage in Argentina and Brazil has doubled from 15 million to 30 million hectares in the last 10 years (FAO 2004a). Despite the transitional nature of the Argentinean and Brazilian economies, the agricultural technology in these countries has been very similar to "developed-world" agriculture: large, mechanized farms growing genetically modified crops. The United States and Europe have indicated their intention to lower farm subsidies, but progress in doing so is slow. In the case of highly subsidized crops like cotton, maize, sugar, and rice, a relatively small number of farmers have been enriched because of powerful political backers. These instances have harmed the U.S. image in other nations and created doubts about whether the political will for reform exists.

In the absence of a serious problem with transgenic crops, it is likely that the total area devoted to these crops will continue to increase in the United States and other countries. Europe will slowly accept transgenic crops, both as imported grain and food and eventually as crops that may be grown by European farmers. Opinions about transgenic crops may also diverge within Europe. More traditional Western Europe may continue to resist the new technology, whereas Central Europeans may become more open to the technology. Though Central European countries are unlikely to push for the adoption of transgenic crops immediately following their recent entry into the European Union, their interests may make them more open over time. Central European countries have traditionally been willing to adopt farming technology that fits well with their open land and large operations.

The disagreement of the United States and Europe over transgenic crops is also reflected in developing countries. Some African countries have refused U.S. food aid and oppose the introduction of transgenic crops because they fear being cut out of the European export market. Pinstrup-Andersen, winner of the World Food Prize, has argued that developing countries should be able to make their own decisions about the new technology, but they are hindered from doing so because of a lack of research capacity and agricultural infrastructure as well as agricultural policy. Underlying the current disagreements is the assumption by some that despite the narrow range of current transgenic products, these traits will eventually impact productivity and cost of the food produced. If this increase in productivity or quality occurs, a country where farmers adopt the technology will be more competitive in increasingly open world markets.

Trade barriers have been built to the adoption of what may become a proven technology, further skewing the economic relations among countries As confidence grows in the safety of transgenic crops and new, more valuable traits are introduced, the debate and its impact may shift from being about safety and environmental concerns to the restriction of trade.

Terrorists' Use of Agricultural Biotechnology

Most studies by the USDA, the National Academy of Sciences, and others have concentrated on terrorists' threats to animal agriculture in the form of animal diseases such as foot and mouth disease (USDA 2003; NRC 2002a). Threats to crop agriculture have been recognized as well. In a white paper by the American Phytopathological Society (APS), preparedness was identified as a primary need. The ability to detect a widespread crop-disease outbreak, diagnose the disease, and move toward recovery are all required to mitigate the threat (APS 2003). Even before September 11, 2001, the NRC had identified the need for a national plan to deal with widespread plant disease or insect outbreaks, whether introduced by terrorists or natural causes (NRC 2002a).

Genomics tools that allow the rapid diagnosis of diseases would be invaluable in early detection and quick reaction. It is also possible for terrorists to use biotechnology techniques, the so-called dual-use nature, to increase the virulence of diseases that infect crops. Insects could be modified to be more voracious or less susceptible to insecticides. While these modifications are clearly feasible from a technical point of view, they seem less dramatic and more difficult to accomplish than the usual tools of the trade for terrorists. Spreading plant disease or insects is highly dependent on the weather and generally unpredictable. The difficulties of controlling a modified disease organism or insect reduce their attractiveness. However, infection could continue for years after the original introduction (Cameron et al. 2001).

Underlying the possible outbreak of a devastating crop disease, food-borne toxin, or major food-contamination scare is the possibility of widespread alarm and destabilization of confidence in the food system (Casagrande 2000). The size of the agricultural economy in the United States—greater than $900 billion—makes the economic threat large, as is the psychological vulnerability resulting from the loss of confidence in the food system.

Public Acceptance

Repeated surveys of public acceptance of transgenic crops and food show different responses among countries. The United States tends to be more accepting of the technology, although a sizable minority is opposed, while Europeans and the Japanese are much more opposed (Falk et al. 2002). Despite a high level of confusion about both the technology itself—many people falsely believe that plants contain genes only if they are introduced through genetic engineering—and the conventional methods of agriculture, recent problems with the food system, including benzene in water, Mad Cow disease, foot and mouth disease, and food-poisoning problems underlie the concern, particularly in Europe. Resistance to the technology has various roots, including food safety, environmental, social, economic, and philosophical. Continued research on food safety and environmental impacts should better differentiate between risks that need to be addressed and those that are not significant.

Unfortunately for the proponents of the technology, several experiences have increased the concerns of those who are opposed. Preliminary data concluding that transgenic, insect-resistant corn pollen could negatively impact the Monarch butterfly was widely reported. Later, more complete studies that found this impact to be unlikely were not well covered by the media (USDA 2004). Two examples of industry failing to comply with regulations have occurred in the United States. Prodigene's field tests of plants containing genes to produce pharmaceuticals in crops required crop destruction that was not carried out. Starlink (Bt) corn, approved for animal feed but not for human food, was introduced into the food system by Bayer Cropscience. It is widely held that no real damage to people or the environment occurred, but these cases raised public concerns, cost the companies involved a substantial amount in fines and liabilities, and represented a major black eye for the industry.

The current products of biotechnology do benefit consumers through the lower cost of commodities, but in developed countries these costs represent a very small percentage of the consumer's food spending. While many of these traits (see table 1), such as modified fatty acids resulting in more healthy oils or increased antioxidants, will have health benefits, they are unlikely to provoke widespread consumer enthusiasm. Even a trait like allergen-free peanut or wheat would be welcome but may not change the debate. Pressure from opponents has limited the adoption of transgenic crops in the United States. McDonald's, concerned about public opposition to French fries made from transgenic potatoes, stopped the introduction of potatoes engineered to resist Colorado potato beetle. Monsanto has delayed the introduction of herbicide-resistant wheat in the United States because of the large export market and uncertainty about acceptance in Europe and Japan. Crops that are associated with direct food use, like wheat, are being commercialized more slowly. Concern that exports to Japan and Europe will be blocked because of their resistance to genetically modified crops and because of the inability to segregate transgenic from non-transgenic grain in the world's commodity systems has resulted in product introduction being postponed.

Considerable controversy has accompanied the production of drugs in crop plants. Some groups believe that pharmaceuticals should only be produced in plants if they can be kept completely separate from those in the food supply. This complete separation will be accomplished either by physical isolation when the crops are grown, genetic mechanisms so the crops do not produce pollen, or by using crops that are not used for food, such as tobacco.

Concerns about the increasing control of agriculture by multinational corporations may be ameliorated if other more public sources of technology emerge, particularly for the developing world's farmers. Even though some concerns about biotechnology are being addressed, it is likely that a portion of the population will continue to be opposed. The fundamental nature of food and our relationship with the things we eat has a great effect on how people feel about changes in the production of food. A number of groups are concerned about technology in some generic way, so they are unlikely to be satisfied by data showing safety. Given eight years of experience with traits that have been commercialized, barring some type of unforeseen problem, transgenic crops will likely continue to be grown and will remain in the food system. Acreage growth in those countries that have adopted the technology is steadily increasing. The movement of these crops outside the five major countries currently growing transgenic crops will probably be slow. The introduction of traits with more value, particularly in developing countries, may increase the acceptance of the technology. European and Japanese negative attitudes toward agricultural biotechnology overwhelm all of the other challenges for research and innovation in this field at the current time. Some elements of this opposition may be addressed through the development of more adequate regulatory oversight, greater access to intellectual property, and the resolution of trade issues between countries. However, opposition on a deeper and more philosophical basis will be harder to overcome.

There are a small number of people whose opposition to transgenic crops is total and whose methods have turned violent. Destruction of field tests of transgenic crops has been common, particularly in Europe. In several cases, research facilities thought to be involved in agricultural research (although this was not necessarily the case) have been bombed. In the United States, laboratories at the University of Washington, the University of Minnesota, and Michigan State University have been bombed. Fortunately, no one has been injured. These bombings are thought to have been carried out by radical environmental groups.

The Impact of Agriculture on the Environment

Transgenic crops raise the possibility both of environmental damage and of lessening the damage caused by current agricultural practices. The genetic nature of transgenic plants means that these plants will in some cases interbreed with wild relatives of the crop. Both the biological impacts as well as the cultural impacts of this potential interbreeding are of concern. It takes a number of generations to tell if a transgene introduced into a wild species from a transgenic crop will result in a competitive advantage for the wild population and even longer to tell if this competitive advantage results in increased success in the environment. Long-term ecological studies have begun to answer these questions (NRC 2002b). In the meantime, introduction of transgenic crops to areas containing wild relatives should be limited, if not prevented. Native farmers who are used to incorporating new material into their open-pollinated varieties have or will attempt to incorporate transgenes. These native open-pollinated varieties are changing entities and must be collected and preserved at intervals over time, while allowing the ever-changing landraces to remain in the hands of the farmer.

Use of insect-resistant cotton in the United States has resulted in less use of insecticides. Farmers used 1,000 tons (active ingredient) less of insecticides (Gianessi et al. 2002). Use of Bt corn for European corn borer did not replace a chemical input, as farmers did not generally treat for this insect. The recent introduction of Bt corn hybrids that are resistant to corn rootworm should result in decreased insecticide use, since many farmers apply insecticide for rootworm control. The use of Roundup Ready crops has shifted many acres from more toxic herbicides to the relatively less toxic glyphosate, but it has not resulted in decreased herbicide use. Transgenic-disease-resistant crops should also result in decreased fungicide use.

Insect- and herbicide-resistant crops place selection pressure on native insect and weed species that can result in resistance to the insecticidal protein or herbicide. Just as is the case with chemical pesticides and traditional genetic resistance, care must be taken to manage the occurrence of this resistance. Active management programs have been introduced along with transgenic crops and will likely slow or prevent the evolution of resistant pests (NRC 2002b).

Most of the earth's suitable farmland is already in use. Expansion of agriculture to more marginal land will result in problems with erosion, poor fertility, and low yields. Crop-yield increases since the 1960s have produced adequate food to prevent the planting of 300 million to 500 million hectares. During this same period, the number of people who are hungry has been reduced (CGIAR 2001). Runge et al. (2003) believe that "the largest contribution agriculture can make to preserving the world's biodiversity is limiting its encroachment on to ever-more-marginal land." If productivity can be increased in a sustainable way without further damage from chemical use and erosion, the world's increasing population can be fed without additional destruction of biodiversity. If the experience with transgenic crops continues without problems, adoption of the technology could have long-term, positive impacts on crop productivity, pesticide and fertilizer use, and the amount of land required to produce enough food for the world's population. While more widespread implementation of traditional agricultural technologies may be adequate to feed the population until 2020, there is a price to pay in the number of hectares devoted to agriculture and other negative side effects of conventional farming.

Summary

There are currently gaps between the science, which is moving rapidly and poised to deliver a revolution in scientific information, the technology, which is being applied to a narrow range of crops, traits, and countries, and the potential broad application of biotechnology. Many potential traits and increasingly efficient technologies will make biotechnology more usable for the developed and developing world. Research by private companies will address opportunities to apply biotechnology to the problems of farmers, consumers, and industry in the developed world. Smaller-acreage crops will be less likely to be impacted by the available technology.

Many important opportunities for increasing the amount and nutritional value of food in the developing world will be addressed. In some cases, companies may provide genes of value for the developing world, but more often, because of the crops and traits desired, public sources of funding for research will be needed. This technical progress can be leveraged from basic research throughout the world and from more applied research in the developing world.

Scientific capability in agricultural biotechnology is currently dominated by the United States; however, restrictions on public funding and foreign-born researchers may weaken the United States' capability and reduce this dominance. Additionally, increased capability in China and India, driven by returning scientists, the importance of agriculture to their economies, and government research funding may make these countries much more competitive. Lack of scientific capability in the rest of the developing world, particularly Africa, will constrain the region's ability to feed its people. Funding for biotechnology research from local governments, large foundations, CGIAR, and assistance from developed countries could address the needs for improvements in developing-country agriculture. But this type of coordinated, concerted effort seems unlikely at this time.

Intellectual property issues are relatively solvable, but not without better understanding of the law in this area. Public researchers and agricultural-research administrators will need to become much more knowledgeable about the functioning of intellectual property law and address those issues head-on. With a broader understanding of patent law, the ability of public research organizations to protect and profit from their discoveries and still provide output for the common good is possible.

The need to address the risks of a new and developing technology through regulation may be the most challenging factor influencing how agriculture will look in 2020 or 2050. Concerns about safety and environmental impacts will tend to result in more expensive regulatory requirements. This will make applications of biotechnology to the developing world slow and may prevent its use altogether. While stronger regulations will increase public confidence in the long term, in the shorter term they will tend to focus application of the technology on large, commercially grown crops. It is not apparent that this can or should happen differently; however, as increased data and experience concerning transgenics are obtained, there will be an opportunity to relax regulatory requirements so these traits can be used more widely.

Interactions among countries, both to reduce protectionism and increase the efficiency of the world's markets, will improve opportunities for the poor and shift agricultural production to the most efficient locations. A level playing field for technology will also be necessary so farmers in any country can compete successfully in an era of more open trade.

Concerns about agriculture's impact on the environment, and what to do about it, are polarized into those groups and individuals comfortable with the use of various levels of technology and those who prefer to forgo modern production methods. Given the current economic drivers of agriculture and the need to feed 3 billion additional people, improved productivity and reduced pollution will be required. Consideration of all approaches to food production and evaluation of risks to make wise choices about the best approaches are essential. It is unlikely that arbitrary decisions that limit options will serve anyone well. Reasonable discussions about desirability and rate of adoption of new technologies could result in better decisions.

More widespread implementation of traditional technologies, along with increasing, targeted biotechnology intervention, will be adequate to feed the population and can be expected by 2020. The more important question regards the time period from now until 2050, when increased population will require more difficult decisions to be made.

References

APS. 2003. Crop biosecurity: Are we prepared? White paper developed by the American Phytopathological Society. http://www.apsnet.org/media/ps/top.asp.

Atkinson, R.C., R.N. Beachy, G. Conway, F.A. Cordova, M.A. Fox, K.A. Holbrook, D.F. Klessing, R.L. McCormick, P.M. McPherson, H.R. Rawlings III, R. Rapson, L.N. Vanderhoef, J.D. Wiley, and C.E. Young. 2003. Public sector collaboration for agricultural IP management. *Science* 301: 174–175.

Boyer, J.S. 1982. Plant productivity and environment. *Science* 218: 443–448.

Cameron, G., J. Pate, and K.M. Vogel. 2001. Planting fear: How real is the threat of agricultural terrorism? *Bulletin of the Atomic Scientists* 57: 38–44.

Casagrande, R. 2000. Biological terrorism targeted at agriculture: The threat to U.S. national security. *Nonproliferation Review* 7: 92–105.

Cassidy, B., and D. Powell. 2002. Pharmaceuticals from plants: The Prodigene affair. Food Safety Network. http://www.foodsafetynetwork.ca.

Cayford, J. 2004. Breeding sanity into the GM food debate. *Issues in Science and Technology* 20: 49–56.

CIGAR, Consultative Group for International Agricultural Research, Technical Advisory Committee (TAC). 2001. *Environmental impact of the CGIAR: An assessment*. Rome: TAC/Food and Agricultural Organization of the United Nations.

Evans, L.T. 1998. *Feeding the ten billion: Plants and population growth*. Cambridge: Cambridge University Press.

Evenson, R.E., and D. Gollin. 2003. Assessing the impact of the Green Revolution, 1960 to 2000. *Science* 300: 758–762.

Falk, M.C., B.M. Chassy, S.K. Harlander, T.J. Hoban IV, M.N. McGloughlin, and A.R. Akhlaghi. 2002. Food biotechnology: Benefits and concerns. *Journal of Nutrition* 132: 1384–1390.

FAO. 2004a. FAOSTAT: Statistics database last updated May 2004. Rome: Food and Agriculture Organization of the United Nations. Available at http://apps.fao.org.

FAO. 2004b. *The state of food and agriculture, 2003–2004*. Rome: Food and Agriculture Organization of the United Nations.

Gianessi, L.P., S.P. Silvers, S. Sankula, and J.E. Carpenter. 2002. *Plant biotechnology: Current and potential impact for improving pest management in U.S. agriculture: An analysis of 40 case studies*. Washington, D.C.: National Center for Food and Agricultural Policy.

Giddings, G., G. Allison, D. Brooks, and A. Carter. 2000. Transgenic plants as factories for biopharmaceuticals. *Nature Biotechnology* 18: 1151–1155.

Gleick, P.H. 2003. Global freshwater resources: Soft-path solutions for the 21st century. *Science* 302: 1524–1528.

Grossniklaus, U., A. Koltunow, and M. v. L. Campagne. 1998. A bright future for apomixis. *Trends in Plant Science* 3: 415–416.

Harrison, P., and F. Pearce. 2000. *AAAS Atlas of population and environment.* Berkeley: University of California Press.

Hirai, M.K., M. Yano, D.B. Goodenowe, S. Kanaya, T. Kimura, M. Awazuhara, M. Arita, T. Fujiwara, and K. Saito. 2004. Integration of transcriptomics and metabolomics for understanding of global responses to nutritional stress in *Arabidopsis thaliana*. *Proceedings of the National Academy of Sciences* 101: 10205–10210.

Holden, C. 2002. Dead zone grows. *Science* 97: 1119.

Huang, J., C. Pray, S. Rozelle. 2002a. Enhancing the crops to feed the poor. *Nature* 418: 678-684.

Huang, J., S. Rozelle, C. Pray, and Q. Wang. 2002b. Plant biotechnology in China. *Science* 295: 674–677.

Interacademy Council 2004. *Realizing the promise and potential of African agriculture: Science and technology strategies for improving agricultural productivity and food security in Africa*. Amsterdam: Interacademy Council.

James, C. 2003. Preview: Global status of commercialized transgenic crops: 2003. *ISAAA Briefs* No. 30. Ithaca, N.Y.: ISAAA.

Jefferson, R.A. 2001. Transcending transgenics—are there "babies in the bathwater" or is that a dorsal fin. In *Future of food: Biotechnology markets and policies in an international setting*. P.G. Pardey, ed. Washington, D.C.: International Food Policy Research Institute.

King, D.A. 2004. The scientific impact of nations. *Nature* 430: 311–316.

Lele, U.J. 2003. *The CGIAR at 31: An independent meta-evaluation of the Consultative Group on International Agricultural Research*. Washington, D.C.: World Bank.

McElroy, D. 1999. Moving agbiotech downstream. *Nature Biotechnology* 17: 1071–1074.

Madhavan, N. 2004. India targets $10 bln in biotech. Reuters press release. July 11.

Marra, M.C. 2001. Agricultural biotechnology: A critical review of the impact evidence to date. In *The future of food: Biotechnology markets and policies in an international setting*. P.G. Pardey ed. Washington, D.C.: International Food Policy Research Institute.

Mervis, J. 2004. Is the U.S. brain gain faltering? *Science* 304: 1278–1282.

Miller, H.I., and G. Conko. 2004. *The frankenfood myth: How protest and politics threaten the biotech revolution.* Westport, Conn.: Praeger.

National Plant Genome Initiative. 2003. *Progress report: 2003-2008.* Washington, D.C.: National Science and Technology Council. http://www.ostp.gov/NSTC/html/npgi2003/index.htm.

NRC [Committee on Biological Threats to Agricultural Plants and Animals]. 2002a. *Countering agricultural bioterrorism.* Washington, D.C.: National Academies Press.

NRC [Committee on Environmental Impacts Associated with Commercialization of Transgenic Plants]. 2002b. *Environmental effects of transgenic plants: The scope and adequacy of regulation.* Washington, D.C.: National Academies Press.

Nottenburg, C., P.G. Pardey, and B.D. Wright. 2001. Addressing freedom-to-operate questions for international agricultural R&D. In *The future of food: Biotechnology markets and policies in an international setting.* P.G. Pardey ed. Washington, D.C.: International Food Policy Research Institute.

O'Toole, J. C., G.H. Toenniessen, T. Murashige, R.R. Harris, and R.W. Herdt. 2001. The Rockefeller Foundation's international program on rice biotechnology. In *Rice genetics IV: Proceeding of the fourth international rice genetics symposium.* G.S. Khush, D.S. Brar, and B.R. Hardy eds. Los Banos, Philippines: International Rice Research Institute.

Pew Initiative on Food and Biotechnology. 2004. *Issues in the regulation of genetically engineered plants and animals.* Washington, D.C.: Pew Initiative on Food and Biotechnology.

Pimentel, D. 1976. World food crisis: energy and pests. *Bulletin of the Entomological Society of America* 22: 20–26.

Pinstrup-Andersen, P., and E. Schioler. 2001. *Seeds of contention: World hunger and the global controversy over GM crops.* Baltimore, Md.: Johns Hopkins University Press.

Qaim, M., and D. Zilberman. 2003. Yield effects of genetically modified crops in developing countries. *Science* 299: 900–902.

Runge, C.F., B. Senauer, P.G. Pardey, and M.W. Rosegrant. 2003. *Ending hunger in our lifetime: Food security and globalization.* Baltimore, Md.: Johns Hopkins University Press.

Smil, V. 2000. *Feeding the world: A challenge for the twenty-first century.* Cambridge, Mass.: MIT Press.

Somerville, C.R., and D. Bonetta. 2001. Plants as factories for technical materials. *Plant Physiology* 125: 168–171.

Teich, A.H., and K. Koizumi. 2004. An introduction to R&D in the FY 2005 budget. In *Research and development FY 2005*. Washington, D.C.: American Association for the Advancement of Science.

Thayer, A.M. 2001. Life Sciences. *Chemical & Engineering News* 79: 25–36.

Thurow, R., and S. Kilman. 2003. As U.S. food aid enriches farmers, poor nations cry foul. *Wall Street Journal*, September 11.

UN Population Division. 2003. *World population prospects: The 2002 revision: Highlights*. New York: United Nations.

USDA. 2003. *Agricultural Research Service Strategic Plan: 2003–2007*. Washington, D.C.: USDA. http://www.ars.usda.gov/aboutus/docs.htm?docid=1766&page=4.

USDA. 2004. Q&A: Bt corn and monarch butterflies. http://www.ars.usda.gov/is/br/btcorn/.

Varmus, H., R. Klausner, E. Zerhouni, T. Acharya, A.S. Daar, and P.A. Singer. 2003. Grand challenges in global health. *Science* 302: 398–399.

Walmsley, A.M., and C. J. Arntzen. 2000. Plants for delivery of edible vaccines. *Current Opinion in Biotechnology* 11: 126–129.

Yu, J., S. Hu, J. Wang, G. Wong, S. Li, B. Liu, Y. Deng, L. Dai, Y. Zhou, X. Zhang, M. Cao, J. Liu, J. Sun, J. Tang, Y. Chen, X. Huang, W. Lin, C. Ye, W. Tong, L. Cong, J. Geng, Y. Han, L. Li, W. Li, G. Hu, X. Huang, W. Li, J. Li, Z. Liu, L. Li, J. Liu, Q. Qu, J. Liu, L. Li, T. Lu, X. Wang, H. Lu, T. Wu, M. Zhu, P. Ni, H. Han, W. Dong, X. Ren, X. Feng, P. Cui, X. Li, H. Wang, X. Xu, W. Zhai, Z. Xu, J. Zhang, S. He, J. Zhang, J. Xu, K. Zhang, X. Zheng, J. Dong, W. Zeng, L. Tao, J. Ye, J. Tan, X. Ren, X. Chen, J. He, D. Liu, W. Tian, C. Tain, H. Xia, Q. Bao, G. Li, H. Gao, T. Cao, J. Wang, W. Zhao, P. Li, W. Chen, X. Wang, Y. Zhang, Y. Xiong, Z. Li, L. Mao, C. Zhou, Z. Shu, R. Chen, B. Hao, W. Zheng, S. Chen, W. Guo, G. Li, S. Liu, M. Tao, J. Wang, L. Zhu, L. Yuan, and H. Yang. 2002. A draft sequence of the rice genome (Oryza sativa L. ssp. Indica). *Science* 296: 79–92.

About the Author

Anthony Cavalieri joined Pioneer in 1981 as a research plant physiologist. In 1994, he became director of trait and technology and in 1995 vice president. Following the acquisition of Pioneer by DuPont in 1999, he assumed responsibility for the combined biotechnology research as it related to crop improvement efforts. Dr. Cavalieri received his bachelor of science degree from the University of North Carolina and his doctorate in biology from the University of South Carolina. He completed postdoctoral research at the University of Illinois. He is a member of many science organizations and communities such the American Society of Plant Biologists, Sigma Xi, the American Association for the Advancement of Science, the Crop Science Society of America, the Science Center of Iowa, the Nature Conservancy of Iowa, and the American Society of Plant Biologists Foundation.